STRATEGIC STUDIES INSTITUTE

The Strategic Studies Institute (SSI) is part of the U.S. Army War College and is the strategic-level study agent for issues related to national security and military strategy with emphasis on geostrategic analysis.

The mission of SSI is to use independent analysis to conduct strategic studies that develop policy recommendations on:

- Strategy, planning, and policy for joint and combined employment of military forces;

- Regional strategic appraisals;

- The nature of land warfare;

- Matters affecting the Army's future;

- The concepts, philosophy, and theory of strategy; and,

- Other issues of importance to the leadership of the Army.

Studies produced by civilian and military analysts concern topics having strategic implications for the Army, the Department of Defense, and the larger national security community.

In addition to its studies, SSI publishes special reports on topics of special or immediate interest. These include edited proceedings of conferences and topically oriented roundtables, expanded trip reports, and quick-reaction responses to senior Army leaders.

The Institute provides a valuable analytical capability within the Army to address strategic and other issues in support of Army participation in national security policy formulation.

i

Strategic Studies Institute
and
U.S. Army War College Press

THE FUTURE OF AMERICAN LANDPOWER: DOES FORWARD PRESENCE STILL MATTER? THE CASE OF THE ARMY IN THE PACIFIC

John R. Deni

June 2014

*****

Comments pertaining to this report are invited and should be forwarded to: Director, Strategic Studies Institute and U.S. Army War College Press, U.S. Army War College, 47 Ashburn Drive, Carlisle, PA 17013-5010.

*****

*****

*****

The author wishes to thank the anonymous interview subjects on the Army Staff and at U.S. Army Pacific who gave generously of their time. Additionally, the author is indebted to Dr. David Lai and Colonel Mark Hinds of the Strategic Studies Institute (SSI), Dr. Maryanne Kelton of Flinders University, Dr. Carlyle Thayer of the University of New South Wales, and Dr. Nicholas Khoo of the University of Otago for their very helpful comments on an earlier draft of this monograph.

ISBN 1-58487-618-2

FOREWORD

The U.S. Army performs a number of critical missions across the vast Indo-Asia-Pacific region. These include underwriting deterrence, building coalition capability, strengthening institutional capacity among partner defense establishments, maintaining interoperability, promoting military professionalism, building operational access, and conducting humanitarian assistance missions. For many, it may come as a surprise to know that almost all of the many Army activities and events that support these missions outside of Northeast Asia are conducted with U.S. Army forces based in the 50 states, often Alaska and Washington State. The roughly 22,000 U.S. Army Soldiers based in South Korea and Japan are focused largely on deterring North Korea from large-scale aggression, and assuring South Korea and other countries of the steadfastness of Washington's alliance commitment.

The costs associated with supporting the increasingly important array of Army military-to-military activities across the Indo-Asia-Pacific theater with forces based in the 50 states present the Army with a significant dilemma—namely, trying to play its vital role in America's broad strategy toward the theater while conducting a post-war drawdown in an era of constrained fiscal resources. In this monograph, Dr. John R. Deni describes, analyzes, and explains the potential benefits and risks associated with a potential solution to that broad dilemma—a reconfigured Army presence in the Indo-Asia-Pacific region. According to Dr. Deni, the time has come for the U.S. Army to reexamine long-held assumptions and move beyond outmoded paradigms, in part by adjusting the Army's presence in East Asia. In this companion study to his recent monograph examining the future of the Army presence in Europe, Dr. Deni provides evidence to support his conclusion that a reconfigured Army forward presence in the Pacific theater could increase the effectiveness of Army efforts, while also providing efficiency gains over time. In doing so, Dr. Deni makes an important contribution to the debate over the future role, mission, and structure of the Army in the Indo-Asia-Pacific theater and to the manner in which strategic Landpower supports broad U.S. national security goals. For this reason, the Strategic Studies Institute is pleased to offer this monograph as a contribution to the ongoing national discussion on the role of the U.S. Army and the manner in which it can best serve the Nation today and in the future.

DOUGLAS C. LOVELACE, JR.
Director
Strategic Studies Institute and
U.S. Army War College Press

ABOUT THE AUTHOR

JOHN R. DENI joined the Strategic Studies Institute in November 2011 as a Research Professor of Joint, Interagency, Intergovernmental, and Multinational Security Studies. He previously worked for 8 years as a political advisor for senior U.S. military commanders in Europe. Prior to that, he spent 2 years as a strategic planner specializing in the military-to-military relationship between the United States and its European allies. While working for the U.S. military in Europe, Dr. Deni was also an adjunct lecturer at Heidelberg University's Institute for Political Science. There, he taught graduate and undergraduate courses on U.S. foreign and security policy, North Atlantic Treaty Organization (NATO), European security, and alliance theory and practice. Before working in Germany, he spent 7 years in Washington, DC, as a consultant specializing in national security issues for the U.S. Departments of Defense, Energy, and State and has spoken at conferences and symposia throughout Europe and North America. Dr. Deni recently authored the book, *Alliance Management and Maintenance: Restructuring NATO for the 21st Century*, as well as several journal articles. He has published op-eds in major newspapers such as the *Los Angeles Times* and the *Baltimore Sun*. Dr. Deni completed his undergraduate degree in history and international relations at the College of William & Mary and holds an M.A. in U.S. foreign policy at American University in Washington, DC, and a Ph.D. in international affairs from George Washington University.

SUMMARY

The time has come for a reappraisal of the U.S. Army's forward presence in East Asia, given the significantly changed strategic context and the extraordinarily high, recurring costs of deploying U.S. Army forces from the 50 states for increasingly important security cooperation activities across the Indo-Asia-Pacific theater. For economic, political, diplomatic, and military reasons, the Indo-Asia-Pacific theater continues to grow in importance to the United States. As part of a broad, interagency, multifaceted approach, the U.S. military plays a critical role in the rebalancing effort now underway. The U.S. Army in particular has a special role to play in bolstering the defense of allies and the deterrence of aggression, promoting regional security and stability, and ameliorating the growing U.S.-China security dilemma.

In particular, military security cooperation programs are becoming increasingly important for achieving U.S. security goals. These military-to-military programs and activities are designed to shape the security environment; prevent conflict through deterrence, assurance, and transparency; and build operational and tactical interoperability. As wartime requirements decrease in the coming year following the end of extensive American involvement in Afghanistan and as the U.S. military undergoes a dramatic yet historically typical post-war drawdown, security cooperation activities will comprise the primary way in which a leaner U.S. military contributes to broad American national security objectives in the next decade.

However, the U.S. Army today remains hamstrung in its efforts to contribute to those broader security goals in the Indo-Asia-Pacific theater. A dated basing paradigm limits the utility to be gained from the roughly 22,000 U.S. Army Soldiers based in East Asia, and the extraordinarily high transportation costs associated with sending **other** U.S.-based Army forces to conduct security cooperation activities across the vast Indo-Asia-Pacific region limits what the Army can accomplish. If reconfigured, the forward-based Army presence in East Asia could help achieve U.S. objectives more effectively and more efficiently. Effectiveness would be increased through more regular, longer duration engagement with critical allies and partners, including Australia, India, the Philippines, Indonesia, Malaysia, Thailand, and Vietnam, while still maintaining deterrence through punishment on the Korean Peninsula. Efficiency would grow by reducing the recurring transportation costs associated with today's practice of sending U.S.-based units to conduct most exercises and training events across the Indo-Asia-Pacific region.

Changing the U.S. Army's forward posture in East Asia involves overcoming several hurdles. These include the challenge of reassuring South Korea and Japan of the U.S. commitment to their security, even as the number of U.S. Soldiers based in those countries decreases; the difficulty of negotiating status of forces agreements and/or cost mitigation arrangements with potential new host nations like Australia or the Philippines; budgetary challenges in terms of funding any necessary initial infrastructure investments; and the need to allay Chinese fears of containment and encirclement. However, these challenges are not necessarily insurmountable. For instance, countries across the Indo-Asia-Pacific theater, including some that have long viewed the United States with suspicion, are coming to value increasingly the offshore balancing role Washington can play vis-à-vis China. Additionally, the one-time infrastructure investment costs associated with any new U.S. forward presence in the Indo-Asia-Pacific region are likely to be offset over a matter of years by savings gained from reduced transportation costs. Finally, Washington can work to explain to Beijing how a transparently reconfigured U.S. presence in East Asia actually benefits China by acting as a pacifier for the more aggressive impulses of American allies and partners in the region, and by re-assuring leaders in those same countries that as China rises, the United States will remain a steadfast partner. There are no guarantees that the United States will succeed in overcoming all of the potential difficulties associated with a reconfigured Army presence in the Indo-Asia-Pacific region, but to avoid trying would severely limit the effectiveness and the efficiency of the Army's contribution to broader U.S. national security goals.

THE FUTURE OF AMERICAN LANDPOWER: DOES FORWARD PRESENCE STILL MATTER? THE CASE OF THE ARMY IN THE PACIFIC

Introduction.

With the January 2012 release of the Defense Strategic Guidance, the U.S. military has increased the attention it pays to the Pacific theater. Officially titled *Sustaining U.S. Global Leadership: Priorities for 21st Century Defense*, what has come to be known as the Defense Strategic Guidance directed the U.S. military to "rebalance toward the Asia-Pacific region."[1] Despite being heralded by some as a dramatic "pivot," some data points indicate that Washington's rebalancing is actually part of an ongoing evolution versus a revolution in U.S. policy. For example, several changes to the U.S. military posture in the Pacific—such as the U.S. Marine Corps' plan to relocate thousands of Marines from Okinawa to Guam—have been underway for some time.

Other changes have been far more recent though, such as the effort on the part of the U.S. Marine Corps to establish a rotational presence at an Australian training facility in Darwin, and appear directly connected to the guidance issued in January 2012. In any event, all of the U.S. military services have taken their cue from the civilian political leadership in Washington, strengthening, initiating, and/or reinvigorating efforts to engage allies, partners, and others throughout the Indo-Asia-Pacific region.

Among the military services, the U.S. Army has been particularly active. This may come as a surprise to outside observers, especially given the sense in the United States that the Pacific theater, outside the context of the Korean Peninsula, is largely the purview of the U.S. Navy and/or the Air Force. There is certainly some logic to that perception, considering the vast distances involved in traversing the theater, which make the mobility platform-intensive Navy and Air Force perhaps better suited to engaging allies, partners, and others throughout the region. In part, this perception has been reinforced by the services themselves, as well as the U.S. Department of Defense (DoD). For instance, in its 55-year history, U.S. Pacific Command (USPACOM), based in Hawaii, has never been led by an Army four-star general.[2]

Nevertheless, the conventional wisdom that the Indo-Asia-Pacific theater is solely or even mostly the purview of the U.S. Navy or Air Force is somewhat outdated. The Army has been and continues to be a major player in the theater as well, judging from not simply Army-led operations during the Korean and Vietnam Wars or ongoing Army-led ballistic missile defense operations in Japan and Guam, but especially in the military-to-military activities undertaken by the Army throughout the Indo-Asia-Pacific region over the last several decades. These activities have proven critical to building the land force capabilities of countries in South, Southeast, and East Asia and Oceania to promote their own security against internal and external threats, to deal with the aftermath of humanitarian disasters by building institutional capacity, to increase professionalism and respect for civilian authorities within partner militaries, to develop operational and tactical interoperability for military operations ranging from peacekeeping to high intensity combat, to further information sharing, to assure treaty allies, and to achieve other shared objectives. For example, the U.S. Army Pacific—based in Hawaii—conducts roughly 200 partnership and engagement activities annually, including 26 major exercises to build partner capacity and maintain varying levels of partner nation interoperability for bilateral and multilateral coalition operations.

Some of the effects achieved by these partnership and engagement activities, such as treaty ally assurance, can be accomplished by U.S. naval and air forces as well. Other effects, such as building capacity to handle humanitarian response crises beyond the littoral, are accomplished more effectively through Army-to-Army interaction and training. Moreover, land forces in the Indo-Asia-Pacific theater have outsized influence in their respective defense establishments—21 of 27 major partners' defense chiefs are Army officers.

Among the many tools the U.S. Army wields in implementing its part of the broader USPACOM theater strategy are those U.S. Army forces based in Alaska, Hawaii, and Washington State—in total, roughly 65,000 Active-Duty Soldiers—who frequently engage allies, partners, and others throughout the theater. This monograph will focus on the role of the **forward-based** U.S. Army forces in the Indo-Asia-Pacific region—that is, those outside the 50 states. Forward-based forces are a powerful tool in the pursuit of both national military and national security goals for several reasons:

- They are a visible U.S. presence in East, South, and Southeast Asia and Oceania;
- They make tangible the many bilateral U.S. security commitments throughout the Indo-Asia-Pacific region;
- They help ensure operational and other forms of access both where they are based and beyond;

- They help promote interoperability with some of America's most capable military allies and most likely coalition partners; and,
- They help build capability among lesser able states for both regional and local stability and security.

However, as with the outdated notion of the Pacific theater as a Navy- or Air Force-only theater, the Army's forward posture in the Indo-Asia-Pacific region reflects a bygone era. In many respects, the forward posture orientation of the Army today—centered on South Korea and, to a lesser degree, Japan—remains rooted in a rationale that has seen little wholesale reassessment since the end of the Cold War. The Army's presence in South Korea is based on the threat that North Korea has posed in one form or another since the end of the Korean War and on the U.S. concomitant treaty obligations to South Korea. The same is largely true of the Army presence in Japan, which is primarily oriented toward logistical support of forces in South Korea but also grounded in a treaty commitment. If the Army's posture in the Indo-Asia-Pacific region were a blank slate today, it is not entirely clear whether it would be in America's interests to base its forces as they currently are. Certainly American treaty commitments to Japan and South Korea remain as vital today as they were 60 years ago. But given the changing strategic context and the role of the Army in fulfilling American strategy—subjects this monograph will examine further—it is conceivable that the existing Army posture is not as effective or as efficient as it might be and is instead a victim of inertia.

Changing that posture would not be easy politically, without up-front costs, or without risk. However, with the return of America's soft power following the low point of George W. Bush's first term, and with rising regional anxiety over Beijing's increasingly overt ability and willingness to translate its economic power into political muscle, the time may be ripe for a reexamination and a reconceptualization of the Army's forward presence in the Indo-Asia-Pacific region.

This monograph is certainly not the first to address the necessity of assessing and possibly reconceptualizing the U.S. military posture in the Indo-Asia-Pacific region. For example, the Center for Strategic and International Studies recently completed a study on the subject of American forward presence in the Pacific theater, the American Enterprise Institute published a study on transforming the U.S. strategy in Asia that counted modifications to U.S. forward presence among its recommendations, and RAND published a report on the strategic choices facing Washington in terms of overseas presence.[3] However, those efforts did not address specifically or thoroughly the role played by the U.S. Army in the Indo-Asia-Pacific region, focusing largely on the Navy, the Air Force, and the Marine Corps. Older, similar strategies and proposals lacked modern budget austerity contextualization.[4] Hence, a reassessment of the U.S. Army's posture during an era of austerity and budget sequestration seems necessary and appropriate.

An analysis of the Army's posture in the Pacific theater must begin by first addressing the changed strategic context, in order to discern the key factors that justify a reexamination of how the Army's forward presence might be wielded in fulfilling U.S. strategy. Next, this monograph will assess the manner in which the United States has responded to the evolving context, with key changes in its strategy, including emphasizing standoff capability, cutting the total size of the military, and "rebalancing" toward the Indo-Asia-Pacific region. One important manifestation of that rebalancing is the role that the Army's forward-based forces have or could have in securing American interests in the Indo-Asia-Pacific region. Finally, the monograph will examine whether and how changes in the Army's forward posture may make those forces more effective and/or more efficient in achieving American ends and furthering American interests.

Change and Continuity in the Strategic Context.

The international environment facing the United States has changed significantly and in a variety of ways over the last 10-15 years. Three aspects of the current international security context are most salient in Washington. First, and perhaps most obviously, the United States now exists in an era of constrained fiscal resources. Argue as some may over whether the Pentagon's budget is bloated following over a decade of war — and there is some evidence that parts of it are[5] — the fact remains that DoD may have to implement some of most significant across-the-board spending cuts in recent memory.

The sequester agreement that was part of the 2011 debt ceiling deal is the most immediate budgetary challenge facing the Department. The 2013 furlough of civilian employees impacted virtually every DoD function, from maintenance to training to strategic analysis to intelligence assessment. What is perhaps worse, though, from a national security perspective, is the impact on readiness. In early-2013, former Secretary of Defense Leon Panetta argued that:

the Department of Defense is again facing what I believe and what the service chiefs believe and what the Chairman of the Joint Chiefs of Staff believe is the most serious readiness crisis that this country is [going to] confront in over a decade.[6]

In late-March 2013, the DoD announced it would end training for all Army units except those preparing to deploy to Afghanistan. It also announced that the Navy would stand down four wings — the equivalent of roughly 240 aircraft — and that the Air Force would curtail training for nondeployed squadrons.[7] Taken together, this adds up to a military that is not as prepared as it should be to defend the interests and security of the United States and its allies around the world.[8]

In addition to readiness, current operations also face restrictions. Already the Pentagon has announced that it will only deploy one carrier strike force in the Persian Gulf, vice two. In late-February 2013, the U.S. Navy also announced it would cancel or defer six deployments.[9] Sequestration may also prevent the Army from deploying follow-on rotations to Afghanistan, thereby prolonging the deployments of units already in the field.[10]

Beyond the immediate challenges posed by sequestration, all indications are that the defense budget is headed downward over the next decade. From the peak of fiscal year (FY) 2010 — when the defense budget was at its highest point in constant dollars since World War II — the defense budget now faces steady cuts for the foreseeable future. In previous drawdowns, such as those following the Korean War, the Vietnam War, and the Cold War, defense spending was cut an average of 33 percent in constant dollars.[11]

The Barack Obama administration's own projections that accompanied the FY2014 budget submission bear out a continuation of this trend, as shown in Figure 1.[12] Regardless of whether one views that positively or negatively from a normative perspective, the fact is that American national security will face increased risk. That risk may be completely acceptable and manageable, or it may not — much will depend on how the United States wields strategy to mediate between risk and cost.

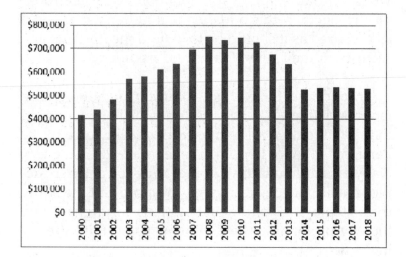

Figure 1. U.S. Defense Spending in Constant Millions of U.S. Dollars.

The second-most important aspect of the current strategic context is a reluctance among senior U.S. leaders to engage in any further land force-intensive operations in Asia or to take on any national security challenge that may require a major reconstruction effort. After a decade of conflict in Iraq and Afghanistan, this sense among senior policy- and decisionmakers reflects public opinion, which long ago began to turn against both wars and, in some ways, reflects the return of the "Vietnam syndrome" of the late-1970s and early-1980s.

Regardless of the argument that the wars and Iraq and Afghanistan were chronically under-resourced, it has become conventional wisdom in Washington, DC, that large land wars, particularly in Asia but also the Middle East or Africa, ought to be avoided.[13] For the national security community, and especially the defense community, this preference among senior decision- and policymakers was manifested through the Defense Strategic Guidance published in January 2012. That document made it clear that the Active-Duty U.S. military force should be, "able to secure territory and populations and facilitate a transition to stable governance on a small scale **for a limited period.**"[14] Not surprisingly, the military services have adjusted accordingly, particularly as they each engage in planning for the post-International Security Assistance Force (ISAF) drawdown of military forces. For example, the Army has eliminated large-scale stability operations as one of the many criteria used to size and structure itself.

Despite the fact that eschewing major counterinsurgency and/or reconstruction efforts overseas, especially in Asia, may be good politics, it is unclear to many in the epistemic community whether it is a realistic policy.[15] Many of those experts are far less sanguine regarding the U.S. ability to pick its enemies and its fights effectively, thereby limiting the conflicts America gets involved in to just those that Washington prefers.[16] Events in Syria over the last 3 years — and the U.S. reluctance to become involved in any extensive way — exemplify the challenges Washington will continue to face in trying to balance the pursuit of its interests against the recent baggage of Iraq and Afghanistan, at least until a new administration takes office in January 2017 and perhaps longer.

More specifically, even as the United States ends large-scale involvement on the ground in Iraq and Afghanistan, maritime-based tensions continue to rise in various locales across the Indo-Asia-Pacific region. While it may be expedient and perhaps accurate to assume that the U.S. Navy would play the lead role at least initially in any American involvement in such a conflict, it is equally certain that as days give way to weeks and months, the United States may be compelled to commit ground forces. Theater and point air and ballistic missile defense, security force assistance, ground surveillance, cyber and network security, and theater sustainment and logistical support are just some of the extensive Army capabilities—in many cases, already in theater—that might be reasonably called upon in the event of a maritime dispute that lasts longer than several days.

Third, there is a growing perception that Asia is increasing in importance when it comes to regions of the world vital to the U.S. economy and hence the American way of life. For example, Asian economies clearly are growing at a faster rate than those elsewhere in the world, even with the recent economic slowdown.[17]

In terms of trade, the picture is somewhat more mixed, at least at first glance. For instance, in 2012, the top 15 U.S. trading partners (imports and exports combined) were Canada, China, Mexico, Japan, Germany, the United Kingdom (UK), South Korea, Brazil, Saudi Arabia, France, Taiwan, the Netherlands, India, Venezuela, and Italy.[18] From a regional perspective, and excluding contiguous neighbors, five of these countries are in East or South Asia, and five are in Europe—a relatively even split. In 2006, by contrast, six were from East or South Asia, and five were from Europe. In 2001, six were from East or South Asia, and six were from Europe.[19] Finally, in 1990, six were from East or South Asia, and six were from Europe.[20] Considered together, this hardly paints a picture of East and South Asia gradually, yet methodically, displacing Europe in terms of importance in trade relations.

A closer examination of the **volume of trade** reveals that Asia is growing in importance to the U.S. economy. In 2001, the dollar value of U.S. trade with the six East or South Asian countries in the top 15 was just over one and a half times that of trade with the top European countries; in 2006, the value of trade with Asia was nearly **twice** that of trade with Europe; and in 2012, the value of trade with Asia was **more than twice** that of trade with Europe.

More specifically, there is also significant evidence that the Indian Ocean has grown in relative importance in terms of global trade.[21] Given the volume of world trade that passes through it each day, the Indian Ocean has become arguably, "the world's most important energy and international trade maritime route."[22] In Australia, Washington's closest ally in the Indo-Asia-Pacific, leaders in Canberra, Australia, see the Indian Ocean as a region of greater importance than even the Pacific or Atlantic Oceans:

> Driven by Asia's economic rise, the Indian Ocean is surpassing the Atlantic and Pacific as the world's busiest and most strategically significant trade corridor. One-third of the world's bulk cargo and around two-thirds of world oil shipments now pass through the Indian Ocean.[23]

China recognizes the growing importance of the Indian Ocean as well, which explains in part its reported search for a series of friendly ports to potentially extend its influence in the region, and its efforts to develop pipelines across Southeast Asia for growing energy demand in southwest China.[24] These and other data points and analyses from both private and public sources appear to support the notion that there is a steady if gradual shift in relative wealth and economic power toward East and South Asia that is likely to endure.[25]

In summary, the United States confronts a strategic context in which resources for national security will remain significantly constrained. With fewer resources at hand, there will be even greater emphasis on preventing conflict and shaping regional and global security environments as efficiently as possible, all in an effort to keep major conflicts and/or massive reconstruction and stability operations at bay. At the same time, the growing importance of the Indo-Pacific-Asia region will compel the United States to reallocate resources from other areas or within the region to make the most efficient use of limited tools in shaping and preventing.

Shifting the Strategy.

Given these and other changes in the strategic context that the United States finds itself in, Washington has responded by shifting its **military** strategy in a variety of ways. First, it has begun to emphasize standoff military capability. This has been exemplified by the new Air-Sea Battle concept and the increased role of drones in U.S. military operations. Additionally, the United States has shown itself increasingly willing to employ—in the right circumstances—an approach that some critics of the Obama administration have derided as "leading from behind." This model—exemplified by U.S. actions in support of several North Atlantic Treaty Organization (NATO) allies involved in the Libyan civil war of 2011—seems a prudent use of American forces and assets, applicable to certain situations where the interests of U.S. allies are vital and American interests are less so.

Second, DoD has begun a significant drawdown of military personnel.[26] From a wartime high of over 560,000 Active-Duty Soldiers, the Army will reduce its end strength to below 490,000 over the next several years. Meanwhile, the Marine Corps is planning on dropping from roughly 205,000 Active-Duty Marines to roughly 182,000. Some argue the cuts should be even deeper.[27] In any case, military leaders hope end strength cuts will result in significant budgetary savings and allow them to protect training and modernization funds.[28] At the same time, military leaders also argue that even though U.S. forces may decline to roughly pre-September 11, 2001 (9/11) levels, the capabilities of tomorrow's military will be far greater than that of the 1990s, given the combat experience of the last decade.[29]

Third, Washington has begun reemphasizing the importance of Asia in its foreign and defense policies. For the DoD, this was most dramatically and most recently expressed in the January 2012 Defense Strategic Guidance mentioned earlier. Since that time, many have interpreted America's "pivot to the Pacific" as a means of containing China with a ring of military alignments, similar in some ways to how the United States sought to contain the Soviet Union through an array of security alliances and agreements—NATO, the Central Treaty Organization, and the Southeast Asia Treaty Organization, for example.[30]

However, U.S. policymakers and senior leaders have gone to great lengths to downplay the role of any containment element in Washington's rebalancing strategy. "Our new strategy and rebalancing in Asia is . . . not about containing China," said General Martin Dempsey, the senior U.S. military officer.[31] More recently, Secretary of State John Kerry appeared to back away from the concept slightly, at least in its military manifestations: "I'm not convinced that increased military ramp-up [in the Asia-Pacific] is critical yet. . . . That's something I'd want to look at very carefully."[32] Just 2 weeks later, former U.S. National Security Advisor Jim Jones characterized the phrase "pivot to Asia" as, "the words we regret most."[33]

Regardless, the "pivot," or rather the rebalancing, is not entirely a new phenomenon. It reflects an evolutionary change — not a revolutionary one — that has been underway for 2 decades as the United States devotes increasing attention to matters in East and South Asia and focuses less on the more limited security challenges in Europe and Latin America. Nonetheless, the Army has responded with additional measures since January 2012, beyond those that have been underway over the last 20 years. For example, the Army has removed the 25th Infantry Division, based in Hawaii, from the pool of forces available for worldwide deployment, which should enable the division to focus more on engaging partner militaries in the Pacific theater. Additionally, the Army has elevated the rank of the USPACOM commander, who is based in Hawaii, from a three-star general officer to a four-star general officer.

Another important element of America's evolving policy toward the Indo-Asia-Pacific theater is the U.S. forward military presence. Whether the United States is rebalancing to engage China or to confront it—among other objectives forward-based military forces can help to achieve—forward presence of U.S. forces plays a vital role. For example, forward presence could be used to contain China through both an intensification of existing bilateral security agreements between the United States and key partners and allies or the initiation of similar arrangements with **new** partners in the region. Alternatively, forward presence could also be used to engage China bilaterally and multilaterally, seeking to build confidence through transparency and the development of mutual understanding much in the way the United States used confidence and security-building measures of the 1980s and 1990s with the former Soviet Union.[34] Actually a third option exists as well—that forward presence could be used simultaneously to achieve **both** of these objectives through a sort of "two-track" approach involving both carrots and sticks.

Because forward presence in the Indo-Asia-Pacific region plays such an important role, it too has been subject to reexamination within the DoD in recent months and years. The unfolding results of this reexamination have included some changes in the U.S. force posture in the Indo-Asia-Pacific region. To a limited degree, these changes have entailed **adding more** to what is already a robust presence in the theater. For example, the recently concluded basing agreement with the Philippines to make use of facilities at Subic Bay and Clark Air Base will provide the U.S. military with an additional location to operate from, as well as a means to engage more actively the Filipino military

15

and develop its capacity to promote security and stability in the region.

In other cases, the changes to posture entail **shifting** assets from others theaters to the Indo-Asia-Pacific region. For example, the U.S. Navy will shift its fleet presence from the current 50-50 split between the Atlantic and the Pacific Oceans, to a roughly 60-40 split favoring the Pacific. The U.S. Navy will also deploy rotationally between two and four littoral combat ships to Singapore.

Similarly, even though the U.S. Marine Corps is reducing end strength overall, it is shifting more of its remaining resources to the Indo-Asia-Pacific region. In 2012, it conducted the first of what are likely to become increasingly larger annual training rotation deployments to Darwin, Australia, through 2016.[35] The first two rotations in 2012 and 2013 included between 200 and 250 Marines, while the 2014 rotation is slated to consist of 1,150 Marines. Eventually, the Marine Corps reportedly plans to send up to 2,500 Marines, as well as fixed and rotary wing aircraft.[36] Elsewhere in Australia, the U.S. Air Force may make greater use of Australian Air Force bases, and the U.S. Navy may conduct more port calls on an Australian naval base outside of Perth.[37]

Meanwhile, the Army presence in the Western Pacific remains seemingly a captive of inertia based on a Cold War paradigm, with Army forces in the region still concentrated in South Korea and, to a lesser degree, Japan. Given strategic and regional dynamics during the Cold War, it made sense to base U.S. Army forces in these locations—after all, the contingency that would most likely require the application of American Landpower in the Indo-Asia-Pacific region was a North Korean invasion of the South. From a practical perspective, there were few other countries interested in permanently hosting U.S. forces in the region.

However, having so many U.S. Army forces tied to missions supporting security and stability on the Korean Peninsula has meant that the Army has had to look elsewhere to source the increasingly important "shaping" and "preventing" activities across the region. It is likely that missions to shape the security environment and prevent conflict will comprise the majority of the U.S. military's post-ISAF activities in the Indo-Asia-Pacific theater over the next decade. Unfortunately, for the U.S. Army to provide personnel and equipment for those missions from locations other than South Korea or Japan is neither an efficient nor effective use of limited resources. For instance, the cost of transporting personnel from Hawaii, Washington State, or Alaska to exercises and training events in the Indo-Asia-Pacific region is extraordinarily expensive, certainly more so than sending personnel from South Korea or Japan, even though that is not an option in most instances today. By one estimate, to send a Stryker battalion's worth of personnel plus a company's worth of their equipment from Washington State or Alaska to the Philippines for an exercise or training event costs roughly double—somewhere between $3 million and $5 million, depending on the amount of advance notice possible—what it would cost to send the same from South Korea.[38] Transportation costs consume so much of U.S. Army Pacific's available security cooperation budget that they are unable to send Stryker Brigade Combat Teams into the Indo-Asia-Pacific region for exercises or training in any significantly meaningful way. Certainly entire brigade combat teams are not necessary for many, if not most, of the security cooperation activities conducted by

U.S. Army Forces, US Pacific Command (USARPAC) around the theater, but especially in cases where advanced operational and tactical interoperability across the range of military operations with America's closest treaty allies is the objective, this necessarily limits the ability of the U.S. Army to contribute to overall American security objectives.

From one perspective, a U.S. Army posture focused on South Korea and Japan may appear to still make sense, considering recent North Korean saber rattling as well as Pyongyang's renunciation of the 1953 armistice agreement that essentially ended the Korean War.[39] However, given the significantly changed strategic context outlined previously and the extraordinarily high, recurring costs of deploying U.S. Army forces from the 50 states for security cooperation activities around the theater, it may be time to reexamine the basis for the Army's presence in East Asia. As argued later in this monograph, the United States may need to consider recasting that presence if it might help to achieve U.S. objectives more effectively and more efficiently without gravely undermining the American commitment to the defense of South Korea and Japan or dramatically worsening relations with China.

The Challenges of Adjusting U.S. Overseas Posture.

Certainly **changing** that presence may not be politically easy, inexpensive, or risk-free. For example, creating forward presence where one does not yet exist is a process usually accompanied by intense, lengthy political negotiations with prospective host nations. However, there is evidence that implies countries of the Indo-Asia-Pacific region may be more amenable to some U.S. military presence today than they have been in some time, even if only rotational in nature. This is because the U.S. ability to be viewed as the security partner of first resort has not been as great as it is today since the opening days of the Cold War, when perceptions of an ideologically and militarily aggressive Soviet Union pushed many in Europe, Asia, and elsewhere — particularly those not occupied by Soviet forces — to seek alignment with America.[40]

Witness, for example, the apparently complete turnaround in attitudes within the Philippines. In the immediate aftermath of the Cold War, political leaders in Manila essentially ejected the United States from military facilities at Subic Bay and Clark Air Base, which were among the largest overseas American bases worldwide. Today however, officials in the Philippines are actively pursuing an American military presence, largely as a means of hedging against growing Chinese influence in the South China Sea, also known as the West Philippine Sea. More specifically, the Chinese government has sent warships to escort large flotillas of fishing boats into the South China Sea to strengthen its claims within the area it has identified by the so-called "nine-dash line," as depicted in Figure 2.[41] Beijing has also accused the Philippines of "illegal occupation" of some of the Spratly Islands.[42] Unable to counter Chinese military power alone and therefore interested in relying on the United States as an off-shore balancer, the Philippines has pursued closer military relations with the United States through basing agreements and an increased program of exercises and training events.[43]

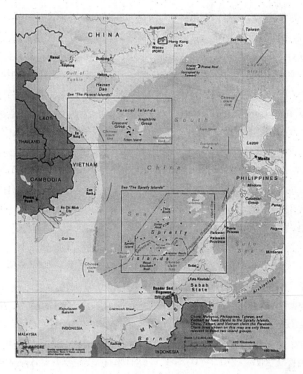

Source: The Perry-Castañeda Library Collection, Austin, TX: University of Texas at Austin.

Figure 2. South China Sea, with the so-called "Nine-Dash Line."

In fact, Manilla is not alone in pursuing such policies. Chinese actions in the South China Sea—as well as historical animosity between the Vietnamese and Chinese—have also been used to explain Vietnam's rapprochement with the United States in recent years.[44] Although reactions to China's growing ambition are not uniform across the region, the combination of domestic politics, changes in the power dynamics of the international system, and historical baggage together have led countries such as the Philippines, Vietnam,

20

Malaysia, and Indonesia to seek closer ties with the United States, while simultaneously balancing their desire for regional and national autonomy.[45] This represents a window of opportunity for the United States, which could leverage regional interest in closer ties to alter and/or expand U.S. forward-based posture in the theater to more efficiently and effectively promote U.S. interests across the Indo-Asia-Pacific region.

Even if political barriers can be overcome and negotiations successfully concluded—as appears to be the case in the Philippines—the budgetary hurdles may prove insurmountable in this era of fiscal austerity, given the up-front costs for even rudimentary facilities and then recurring variable and fixed costs associated with forward presence. For instance, the case of U.S. efforts to create and maintain forward operating sites[46] in Romania and Bulgaria sheds light on the up-front, one-time costs to establish new facilities, as well as the recurring costs of maintaining such facilities. In this example, the United States spent roughly $110 million to develop basic training and life support facilities at existing Romanian and Bulgarian military bases, which together are capable of hosting an American brigade combat team.[47] Since most U.S. forces use those facilities for exercises and training only on a periodic or rotational basis, most of the variable recurring costs are funded through exercise and training programs. Otherwise, the majority of **fixed** recurring operating costs are those associated with the so-called "warm basing" of the site in Romania and the "cold basing" of the site in Bulgaria, which entails maintaining the facilities between rotational deployments and managing any requirements generated by ongoing U.S. humanitarian and civic assistance missions. The warm/cold basing costs for the sites in Romania and Bulgaria are roughly $11 million per year.[48]

This provides a useful data point for having a sense of what changes to overseas posture can cost in terms of both one-time costs and recurring variable and fixed costs. Of course, costs in Bulgaria and Romania are very likely to differ from those in East Asia for any number of reasons, and so a more broadly based study of overseas presence costs can also help inform any assessment of one-time and recurring costs, as well as potential cost mitigation strategies for the United States. Such a cost study, recently directed by the U.S. Congress and completed by RAND, was designed to assess the relative costs and benefits between overseas permanent and rotational basing on the one hand and U.S. basing on the other. The RAND study found, perhaps unsurprisingly, that the largest single cost driver in switching from the current posture to an alternative posture was the cost of new construction.[49] In arriving at this conclusion, the study authors assumed that the "alternative" would involve relocating forces from outside the United States to facilities inside the United States, but given the other conclusions reached by the study authors discussed here it seems reasonable to assume that the same would apply to a relocation from one overseas location to **another** overseas location as well. In terms of cost mitigation strategies, if the United States were to shift Army forces within the Indo-Asia-Pacific theater on either a permanent or rotational basis, it would be well-served by concentrating only on those locations with the most advanced pre-existing life support and training facilities, thereby minimizing the costs of any necessary upgrades.

Regarding whether an alternative **rotational** presence might be cheaper to maintain than an alternative **permanent** presence, somewhat counterintuitively to those unfamiliar with overseas presence issues, the study found that, "rotational presence is not necessarily less expensive than permanent presence."[50] More specifically, the study concluded that costs associated with rotational basing depend on the frequency and duration of deployments.[51] Higher frequency deployments of shorter duration generated more cost, travel costs in particular, than lower frequency deployments of longer duration. Hence, if the United States cannot sustain the costs associated with permanent overseas presence, a rotational presence of long-duration deployments of many months, or even a year, would likely prove less expensive, and hence more sustainable from both budgetary and political perspectives, than short-term deployments of only several weeks or a few months.

Of the costs associated with long-duration, less frequent rotational deployments overseas, the congressionally mandated study also found that personnel-related costs, such as food, housing, and special allowances, were the most significant drivers. The United States may be able to mitigate some of these major costs by negotiating for prospective host nations to assume or at least share housing and food costs. Alternatively, special personnel pay allowances seem an unlikely candidate for cost-sharing arrangements.

When it comes to negotiations over cost sharing, the United States probably will not be able to replicate the direct support arrangements that are in place today with Japan, which has helped to offset by roughly 75 percent the costs of permanently basing U.S. forces there.[52] Nonetheless, even here the new strategic context may prove a benefit to the United States, forcing a new strategic calculus on countries in the Indo-Asia-Pacific region and potentially making them more amenable to co-funding, defraying, or otherwise offsetting the costs of an American military presence. For example, Germany provides tax and customs relief for U.S. forces, rent-free property for U.S. basing, thousands of acres of rent-free military training land, and other contributions that have offset roughly one-third of U.S. basing costs.[53] Further south, Spain has provided similar indirect support that has offset more than half of U.S. basing costs.[54] Similar offsetting, indirect support arrangements in the Indo-Asia-Pacific region would go far in reducing U.S. forward presence costs, while enabling prospective host nations to avoid the appearance of outsourcing national defense by providing direct payments for U.S. forces.

Finally, in addition to the hurdles described previously, changing the Army's forward posture may have the unintended consequence of engendering a sense of encirclement or alienation among the Chinese. The roots of China's strategic distrust of the United States extend at least to the founding of the People's Republic of China in 1949—since then, the Chinese have largely maintained wariness toward America. U.S. efforts to shift more attention toward the Indo-Asia-Pacific region, including as promulgated through the January 2012 Defense Planning Guidance, appear to be reinforcing those Chinese sentiments.[55] For example, Washington's strengthening of security ties with countries such as India and Vietnam—two countries that have fought border wars with China and that have not been traditional U.S. allies or even partners in most instances—have caused many in Beijing to believe that the United States is bent on containing China. More broadly, Chinese leaders have come to view **American** policies, attitudes, and misperceptions as the chief cause of strategic mistrust between the two countries.[56] Even U.S. efforts to promote

24

democracy and human rights are interpreted in Beijing as nothing more than Trojan horses for the expansion of American power abroad, often at China's expense.[57] Hence, although the United States has arguably done more than any other country to contribute to China's ongoing modernization, America's rebalancing risks feeding China's sense of encirclement, undermining regional stability, and decreasing the possibility of cooperation between Beijing and Washington.[58]

However, Beijing may also come to understand that a different distribution of the Army presence in the Pacific theater — or even an increase in that presence — may actually benefit China in two distinct ways. First, a redistributed American Army forward presence may act as a pacifier for some of the more aggressive tendencies of U.S. allies in the region. For instance, given recent changes in the security dynamics of the region, the Philippines announced in mid-2013 that it is spending $1.8 billion to expand and modernize its military aggressively in order to counter Chinese influence.[59] A redistributed U.S. Army presence in the Indo-Asia-Pacific region might better enable military and security ties between the United States and its allies, such as the Philippines, thereby expanding American influence and social access and enabling the United States to better play the role of strategic pacifier with those allies and partners. This would be similar in some ways to how the former Soviet Union came to agree with the U.S. proposal to keep a unified Germany in NATO following the end of the Cold War. The Soviet Union came to see that it would be more secure with a unified Germany under the American security umbrella rather than a neutral Germany untethered from NATO and free to exercise its power in a more unilateral fashion.[60]

Second, a redistributed U.S. Army presence in the Pacific may make American allies throughout the theater more comfortable with a rising China, which, of course, would benefit Beijing as well. The presence of U.S. Army forces represents a tangible sign of American commitment on the ground to allied security. Building and maintaining the confidence of U.S. allies in the steadfastness of the American commitment to their security can benefit China by potentially mitigating some of the worst fears in the Philippines, Thailand, Australia, and elsewhere over China's rise. This is similar in some ways to how Germany has supported continued American military presence in central Europe. Such support from Berlin is not based simply on the local impact of American dollars being spent in rural areas of Germany, where U.S. military facilities tend to be located, but also on a shrewd German understanding of how U.S. troop presence in Germany helps reassure Poland, the Czech Republic, France, and others regarding a growing, unified Germany as the most powerful country in Europe. Hence, in the contemporary Pacific context, it is plausible that China might actually, although perhaps not overtly, acquiesce to a changed disposition of Army forces in the Indo-Asia-Pacific region.[61]

Convincing Chinese officials of the potentially beneficial aspects of a redistributed or increased U.S. Army presence in the Indo-Asia-Pacific will not be easy—some in Beijing suspect Washington is, in fact, actively fomenting aggressive behavior on the part of U.S. allies in the region.[62] But there is already some limited evidence that China might welcome a pacifying role played by the United States vis-à-vis aggressive tendencies of American allies. For instance, an editorial in *China Daily*, which tends to reflect official opinion in China, noted that with regard to the dispute with the Philippines over competing island claims:

> The joint defense between the United States and the Philippines . . . will be a favorable factor to stabilize the situation in the South China Sea as long as the United States insists on joint defense and opposes to joint infringement and external expansion by its ally.[63]

Elsewhere, some Chinese scholars and officials seem open to the notion of recrafting the great power relationship between China and the United States.[64] Part of that redefinition could include a recognition within China of the important role the United States can play in helping to ameliorate security dilemmas in the Indo-Asia-Pacific. Nonetheless, whether Washington can succeed in convincing the government in Beijing of this—or whether the United States is willing to take other steps that China believes are necessary for amelioration of the security dilemma—remains to be seen. Regardless, whether Washington seeks to challenge and contain China, or to engage her, or even to do both, it is worth investigating whether the U.S. Army can contribute more effectively and more efficiently to American national security by moving beyond the Cold War paradigm that has led to an emphasis on the Korean Peninsula.

The Rationale for a Reassessment.

American military forces are forward-based for several reasons. First and foremost, the United States bases military forces abroad to defend vital U.S. interests by safeguarding the security of its most important allies. Examples might include the American presence in Germany during the Cold War or the American presence in South Korea today. Additionally, U.S. forces may be based overseas to be an ocean closer to critical lines of communication necessary for American security or that of its allies. The U.S. presence in Iceland during the Cold War was an example of this, promoting Washington's ability to secure and defend the North Atlantic approaches to both North America and Europe.

Forward presence also contributes directly to building and maintaining interoperability with America's most likely, most capable coalition partners, and to building and maintaining more limited but no less important capabilities among other, less capable partners. Examples of this include the U.S. presence in Europe today, which is critical to maintaining interoperability with highly capable allies such as France, Germany, Israel, Italy, and the United Kingdom, and to building the capability of militaries in southeastern Europe, North Africa, and Sub-Saharan Africa to promote stability and security internally and in their immediate vicinities.

Finally, forward-based U.S. forces provide logistical support to American and allied forces elsewhere. Examples of this can be seen in elements of the U.S. presence in Germany and Japan today. During the last decade or more of war, the American presence in and around Kaiserslautern in southwestern Germany — home to Ramstein Air Base and the Landstuhl Regional Medical Center — has been critical to sustaining operations in both Iraq and Afghanistan. Likewise, in the Pacific theater, U.S. Army forces in Japan provide logistical support to forces in South Korea.

The United States has maintained forward-based military forces in East Asia since the early-20th century for many of the reasons cited previously. However, the origins of today's U.S. troop presence in East Asia are found in the broader Cold War era effort to provide a bulwark against systemic, if not existential, communist aggression. For the Army, this was most evident on the Korean Peninsula and in Southeast Asia, where American Landpower faced significant tests during the Korean War and the Vietnam War.

On the Korean Peninsula today, the deterrent role of the U.S. Army remains important. The presence of over 19,000 Soldiers in South Korea provides a tangible manifestation of U.S. support for South Korea, which would be important in any immediate response to a North Korean invasion and would act as a trip-wire, compelling an even greater U.S. response in the event of large scale hostilities initiated by Pyongyang.

Nonetheless, it appears as if the United States is preparing to take something of a backseat in the defense of the Korean Peninsula. As part of the Strategic Alliance 2015 agreement signed in July 2010, the South Korean military is soon to comprise the "supported" force, with the U.S. military playing a backup or supporting role. The United States will transfer wartime operational control from the American-led U.S.-South Korean Combined Forces Command—which will be disestablished—to the South Korean Joint Chiefs of Staff by the end of 2015, which will give South Korea primary responsibility for leading a response to any North Korean incursion while maintaining the U.S. commitment to South Korea's defense. Although this date may slip somewhat due to concerns over whether and when South Korean forces will be ready, the underlying plan to transfer control has not been questioned by the United States or South Korea.

Additionally, U.S. forces will consolidate and relocate from bases around Seoul to more centralized locations south of the city—this will "improve efficiency, reduce costs, and enhance force protection by placing most service members and equipment outside the effective range of North Korean artillery."[65]

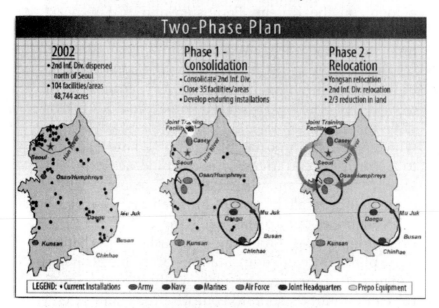

Figure 3. U.S. Military Plan for Consolidation and Relocation in South Korea.[66]

For their part, the South Koreans appear to be committed to the task of assuming wartime operational control of their own forces, declaring their intent to strike back at the North if attacked.[67] Moreover, U.S. Forces Korea—the American military command in South Korea—characterizes the South Korean military as "one of the most progressive and efficient defense organizations in the world."[68] At the same time, and in part due to demographic challenges, the South Korean military is in the process of reform designed to increase the qualitative capabilities of the armed forces at the expense of quantity—as part of Seoul's

Defense Reform Plan 2020, the Active-Duty military force will be cut by roughly 27 percent from 655,000 troops to roughly 500,000, but the defense budget will increase so as to modernize equipment and develop a more professional military force.[69]

The relocation of U.S. forces from the inter-Korean border to south of Seoul should help increase the odds of their survivability in the event of any North Korean attack, as well as reduce the costs of keeping U.S. forces in South Korea. However, with Americans no longer on the front lines, literally and figuratively given the impending South Korean assumption of wartime operational control, this begs the question of whether and to what degree the U.S. Army presence in South Korea deters the North. Although difficult to assess given the opacity of the North Korean regime, it would appear at first glance that the deterrent value of U.S. Army forces relocated south of Seoul and no longer in operational lead would be more limited than a generation ago, when U.S. Army forces were more numerous in South Korea,[70] more closely located near the demilitarized zone, and in the lead role. Nevertheless, upon further examination, it seems that although U.S. forces may contribute less today to deterrence by **denial**, they certainly or at least evidently still contribute to deterrence by **punishment**.[71] This appears to be the case because despite the reconfigured, relocated U.S. Army forces in South Korea, the North Koreans have not initiated a major attack. Certainly, there have been hundreds, even thousands, of North Korean violations of the armistice since 1953, most of them minor, but the long-feared artillery barrage of Seoul and subsequent massive armored and infantry invasion of the South has not occurred. The regime in Pyongyang evidently appears assured that the remaining, reconfigured U.S. forces in the South continue to play at least a "tripwire" role—in other words, the North Korean regime seems convinced that there would be an overwhelming American response to major aggression.

Nonetheless, what remains unclear is what size of a tripwire is necessary. If U.S. Army forces in South Korea are to play any role beyond that of tripwire, the most useful role to play would likely be one the South Koreans cannot perform themselves or cannot perform very well. However, and as noted previously, the South Korean military is perhaps more capable and full-spectrum than at any time in its history and likely to become more so, thanks to ongoing reform efforts. Regardless, and perhaps most importantly for stability and security in Northeast Asia, any further reconfiguration of the American presence in South Korea must be accompanied by overt or explicit signaling, statements, and other actions by the United States of its resolve to remain steadfast in its alliance with the South and to punish any major aggression from the North.

Meanwhile, in Japan, which hosts the second largest concentration of U.S. Army forces in theater—roughly 2,600 Soldiers—the government in Tokyo appears uninterested in taking up any mantle of global ambition. Indeed, Japanese Self-Defense Forces (SDF) remain hobbled by what some analysts call "anachronistic constraints,"[72] which prevent them from engaging more energetically overseas. For this reason, interoperability between U.S. Army and Japanese Ground SDF has been limited. Some have blamed this on a contrast in focus, with the U.S. Army emphasizing wars in Southwest and Central Asia, while Japanese land forces have concentrated on peacekeeping operations and disaster relief operations.[73] More broadly though, Japan's Ground SDF has never emphasized deployability or expeditionary, amphibious capabilities for reasons largely cultural and constitutional.[74]

The U.S. Army's mission in Japan today is to operate port facilities and a series of logistics installations throughout Honshu and Okinawa. Key units based there include a combat sustainment and support battalion, an aviation battalion consisting of UH-60 helicopters and UC-35 fixed-wing aircraft, a signal battalion, an ordnance battalion, and a military police detachment. Meanwhile, 15,300 Marines, 12,700 Airmen, and 3,400 Sailors are also based in Japan.[75] If the U.S. Army mission in South Korea were modified and the resulting forward presence there re-vamped in accordance with the modified mission, the Army may find decreasing utility for its forces in Japan and might consider turning over residual logistical functions to the Navy, which has a far larger presence in Japan and some of the very same logistical capabilities there.[76]

What Might a Reconfigured Army Presence in Asia Look Like?

If the U.S. Army's mission in Northeast Asia were modified so as to enable the Army to achieve its objectives with fewer Soldiers in both South Korea and Japan, it is conceivable that the Army could make a greater contribution to theater-wide U.S. national military and national security objectives.[77] It is even conceivable that such a contribution might be made more efficiently and more effectively than is presently the case.

The strategy of the USARPAC—the Army component command within the larger USPACOM—centers on achieving five related objectives across the Indo-Asia-Pacific region.[78] The first of these—the highest priority—centers on assuring treaty allies of the steadfastness of the U.S. defense commitment and on maintaining and expanding operational and tactical interoperability. Of the six treaty allies in the Indo-Asia-Pacific theater, South Korea is arguably the only one where the physical presence of U.S. Army Soldiers makes a tangible, necessary difference in the deterrence of aggression against that treaty ally. None of the others—Australia, Japan, the Philippines, and Thailand—face the same type of threat for which a U.S. Army physical presence forms a necessary policy tool in terms of "assurance," although as argued later in this monograph, the imperative to maintain interoperability for the most complex military operations across the region and beyond with a country such as Australia may form the basis for a reconsideration of this conventional wisdom.

Whether the U.S. Army presence in South Korea—the tripwire that enables deterrence by punishment—needs to remain as high as 19,000 Soldiers while still assuring Seoul and South Korea's neighbors of the American commitment is open to debate. Arguably, that presence could be reduced to center on air and missile defense units and military intelligence, surveillance, and reconnaissance units—two functional areas that even the increasingly capable South Korean military would likely garner great value from continuing to host and interact with—while still providing South Korea with treaty-based assurance and the ability to conduct bilateral exercises and training events. Additionally, shifting U.S. Army resources from South Korea and Japan to elsewhere would likely increase their operational resilience by dispersing them around the Indo-Asia-Pacific theater. At present, most of the U.S. Army's "eggs" are in two baskets, South Korea and Japan, arguably the highest threat zones throughout the entire theater. This places them at great risk of strategic failure in the face of increasing threats from precision-guided munitions. Finally, a reduction in the Army presence in South Korea would have the added benefit of reducing administrative and logistical costs to the Army. Because most U.S. troops serve in South Korea on 1-year tours of duty, unaccompanied by family members or other dependents, roughly 600 to 700 U.S. servicemen and servicewomen arrive or leave South Korea **each month.**[79] Indeed, the current commander of U.S. Forces Korea, General James Thurman, has asked the Army and other U.S. military services to investigate how they might collectively mitigate the negative effects created by the high turnover rate.[80]

If the number of U.S. Soldiers in South Korea were reduced—as well as the number of U.S. Soldiers in Japan—and repositioned elsewhere in the theater on a permanent or long-duration rotational basis, those U.S. forces could be utilized to achieve USARPAC's theater-wide objectives far more cheaply than flying in personnel and equipment from Hawaii, Alaska, or the continental United States. After "assure," the next highest priority for USARPAC is to promote the ability of key partners to assume greater responsibility for security and stability in their region and beyond. This essentially entails developing among key partners their ability to lead and participate in multinational crisis response operations across the range of military operations, with or without U.S. forces.

If such operations are to occur in conjunction with U.S. forces, then interoperability is critical—namely, the ability of allied or partner military forces and those of the United States to operate side by side or embedded with each other in military operations. America's most important ally in the Indo-Asia-Pacific region in this regard is Australia, with good reason. In repeated instances, the Australian government has proven itself willing and able to deploy and sustain Australian military forces for operations with U.S. forces. Afghanistan provided the most recent evidence of this, and there is no reason to think this will change any time soon, despite a period of budget austerity on both sides of the Pacific.[81] The commonality of world outlook; the shared values; and the professional, capable, expeditionary nature of the Australian military make Australia one of America's most likely, most capable future coalition partners.[82]

Most recently, U.S. Army and Australian forces worked together in Uruzghan province in Afghanistan, building interoperability on the ground in the process. With the ISAF mission in Afghanistan winding down over the next 1 1/2 years and with Australian forces poised to end their mission in Afghanistan by late-2013, there is no Army presence in Australia today or planned for tomorrow that might be used to maintain interoperability with this critical ally. Moreover, the ABCA program—designed to promote interoperability between the armies of America, Britain, Canada, Australia, and New Zealand—has languished in recent years due to a lack of available forces, funding cuts, and episodic senior-level attention within the Army.[83] When U.S. Army forces **are** sent to exercise with Australian counterparts—for instance, as part of the Talisman Saber exercise series, held for 3 weeks every other year—they are typically deployed at significant cost from the United States, particularly the 25th Infantry Division based in both Hawaii and Alaska, as well as Army units based in Washington State.

Although the United States lacks the same kind of strategic interoperability imperative with India that it has with Australia, it remains nonetheless important to Washington to promote the ability of India to exercise regional, if not global, leadership in promoting security and stability. Deepening defense and security cooperation with India is therefore of critical importance to Washington, particularly as ISAF and the large American presence in Central Asia comes to an end.[84] In fact, former Secretary of Defense Panetta called defense cooperation with India "a lynchpin" of the U.S. strategy to rebalance to the Indo-Asia-Pacific region.[85] To this end, the U.S. Army has responded by intensifying its relationship with the Indian Army, focusing specifically on building capabilities for peacekeeping operations. In 2004, the two armies began conducting an annual training event dubbed "Yudh Abhyas," the first such engagements since 1962. In 2009, the series expanded from what were relatively small annual exchanges focused on command post activities—that year, roughly 250 U.S. Soldiers based in Hawaii traveled to India, with 17 Stryker vehicles, to take part in live-fire and field training exercises over 2 1/2 weeks. In 2010, roughly 150 Indian soldiers traveled to Joint Base Elmendorf-Richardson in Alaska for 2 weeks to participate in Yudh Abhyas 2010. In March 2012, 170 U.S. Soldiers based in Hawaii, Alaska, and Japan traveled to an Indian Army training area in Rajasthan for 2 weeks to conduct Yudh Abhyas 2011-12, a live-fire event and a field training exercise built around a peacekeeping operation scenario. The 2011-12 iteration was noteworthy for the deployment of three U.S. tanks to India to participate in the event. In May 2013, roughly 200 Indian soldiers traveled to Ft. Bragg, NC, to participate in Yudh Abhyas 2013 for 2 weeks, which included a combined airborne operation.

If some of the U.S. Army forces in South Korea and Japan were repositioned closer to Southeast Asia, security cooperation events designed to maintain strategic, operational, and tactical interoperability with the Australian Army for the full range of military operations and to build operational and tactical interoperability with the Indian Army for a more limited set of operations, including peacekeeping, potentially could be executed more easily, cheaply, and frequently — transportation costs alone would be cut at least in half, compared to sending forces from the 50 states to participate in such events.[86] At the same time, those U.S. Army forces could more easily and cheaply also accomplish the third USARPAC objective — enhancing critical capabilities among partners so that those partners can address internal security challenges and participate in regional peacekeeping, humanitarian assistance, and disaster relief operations.

Enhancing the capabilities of partner militaries in this way is an objective pursued by USARPAC across the entire Indo-Asia-Pacific region, including China specifically. Obviously, some countries are more willing than others to participate in the kinds of security cooperation events that help to achieve these objectives, such as small-scale bilateral exercises and subject matter expert exchanges. Nevertheless, as evidenced by the cost data discussed earlier in this monograph, the cost of engaging willing partners to achieve U.S. objectives would likely be lower if the Army had forward-based forces repositioned closer to Southeast

Asia, as well as those remaining in Northeast Asia, vice deploying forces from Hawaii, Alaska, or the continental United States.

Repositioning U.S. Army forces toward Southeast Asia would also enable USARPAC to achieve its fourth and fifth priorities more effectively in the Indo-Asia-Pacific theater—opening new relationships and sustaining traditional, less intensive relationships—but the efficiency gains would likely be more limited than those described previously. These last two objectives are pursued largely through senior-leader engagements and small group exchanges, usually involving limited numbers of personnel and little to no military equipment.

Nonetheless, it seems that there may indeed be efficiencies to be gained and costs to be reduced through a repositioning of some U.S. Army forces toward Southeast Asia. When it comes to specific locations where U.S. Army forces might be relocated **to** from South Korea and/or Japan, or whether such a repositioned forward presence might be permanent or rotational, the options depend first on the objectives to be achieved. For instance, the Army's experience both in Europe and South Korea has shown that, in order to maintain ground forces strategic, operational, and tactical interoperability across the full range of military operations, permanent forward presence is preferable to rotational presence.[87]

Given Australia's role as the closest, most interoperable, most capable ally in the theater today and the fact that Australia is likely to remain America's most capable, most likely future coalition partner, it seems logical to consider a permanent U.S. Army presence there, or at least long-duration rotational deployments. As noted previously, the Marines have already begun short-term rotational deployments to an Australian military training facility near Darwin, on the northern coast of Australia. The U.S. Army needs to be interoperable with the Australians, too, particularly as the Marine Corps returns from essentially functioning as America's second land force in Iraq and Afghanistan to being the light, forced-entry military component, which is their comparative advantage. So far, public opinion surveys in Australia reveal growing support for the Marines' rotational presence—in 2011, 55 percent of Australians were either strongly in favor or somewhat in favor of the U.S. presence in Darwin, whereas in 2013, that figure had risen to 61 percent.[88]

A mid-sized city of 130,000 people with a history of hosting military service members, Darwin represents a potential location for an initially small presence of U.S. Soldiers—ideally permanent, but perhaps consisting of medium- to longer-term rotational deployments at first. Such a presence could be usefully modeled on the Bulgaria/Romania example cited earlier, which would focus on extensive interoperability training with nearby Australian forces but also include shorter-duration training and exercise deployments from the warm base in Australia to Malaysia, Indonesia, Thailand, India, Vietnam, the Philippines, and elsewhere. With an English-speaking population and a Western culture, Australia provides a potentially attractive host nation environment, as the U.S. Marine Corps is now learning. The Australian Army's Robertson Barracks—20 minutes' drive from the center of Darwin—plays host to a light-armored brigade, which trains at the Bradshaw Field Training Area, roughly 600 kilometers southwest. The Bradshaw Field Training Area is a 2.1 million acre training site, which is the largest in Australia and roughly three times the size of the National Training Center at Fort Irwin, CA. Bradshaw Field includes a joint and combined Australian-U.S. training center, as well as an airstrip capable of handling C-17 *Globemasters*. However, other infrastructure at Bradshaw Field is quite limited, and excessive precipitation during the rainy season makes it difficult to fully utilize the training area from late-November until early-April.

Whether a cost-sharing agreement could be reached with Australia for a permanent or long-duration rotational presence at Robertson Barracks or Bradshaw Field, of course, remains to be seen. However, with an arrangement similar to what U.S. forces enjoy in Germany—including no-cost lease of land, no taxes, no customs duties, and cost-free access to existing military training grounds—the Australian government thereby could indirectly carry roughly 30 percent of the cost of basing U.S. Soldiers there. From the U.S. Army's perspective, the roughly $21 million that is spent on theater security cooperation events across the Indo-Asia-Pacific theater every year—the vast majority of which is spent on transportation of personnel and equipment participating in roughly 14 exercises annually—could be spread among many more events if the transportation-related cost of participation were cut by 50-75 percent.

A second possible location for a more robust U.S. Army presence may be the Philippines. At present, though, media reports indicate the Filipino government has agreed only to a rotational presence of U.S. ships, aircraft, and troops for training events, exercises, and disaster and relief operations, with the potential of pre-positioning military equipment as well.[89] In any case, an initial rotational presence of increasingly longer durations may provide a starting point for consideration of other, varied U.S. Army forward presence options in the longer-term.[90]

Given its central location in the Indo-Asia-Pacific region, relative at least to Australia and certainly to South Korea, transportation costs associated with a robust security cooperation program—increasingly critical to American national security as described in the strategic context section previously—would most likely be far less than they are today. Moreover, response times to potential crisis locations across the increasingly critical lines of communication from the Indian Ocean to the Pacific would be shorter. Additionally, space for an increased American presence—whether permanent or rotational—exists in the form of what remains of the shuttered Subic Bay naval facility and Clark Air Base, which were the largest overseas American military facilities in the world during the Cold War period. Given the pressure Chinese actions in the South China Sea are generating in Manila, the government there may be amenable to some cost-sharing arrangement in what has been described as a "renaissance" period in U.S.-Filipino relations.[91]

A third potential location for an increased Army presence in Southeast Asia may be Thailand. The U-Tapao Royal Thai Navy Airfield is currently the only facility in Southeast Asia capable of supporting large-scale logistical operations.[92] Although the United States maintains significant military, intelligence, and law enforcement ties with Thailand, it is unclear whether the government in Bangkok today has the same strategic outlook and interests as the United States. Hence, its "reliability as a partner and its ability to be a regional leader" are uncertain."[93] More specifically, the 2006 military coup has complicated U.S. relations with Thailand, and the country is therefore a less attractive partner for a permanent or even rotational U.S. Army presence.[94]

Regardless of the specific location—Thailand, the Philippines, Australia, or elsewhere—Washington would need to carefully manage the presence, activities, and numbers of permanent or rotationally deployed U.S. Army forces. American military activities overtly aimed against China or Chinese interests would likely magnify the sense among potential host nations that they are choosing between the United States and China, a position that is anathema for most countries of the region. Clear benefits to local populations—for example, in terms of contracts with local businesses, employment of local nationals, and engagement with and support for local civic organizations—would go far in solidifying positive perceptions of any new American military presence in Southeast Asia.

Conclusion.

In a period of declining defense budgets and decreasing military end strength, the U.S. military cannot afford to continue to add tasks to its already lengthy to-do list. Likewise, "doing more with less" is a recipe for strategic and tactical failure. In the specific context of America's rebalance to the Indo-Asia-Pacific region, characterized previously as an unfolding evolution in American security policy, it appears increasingly evident that, to achieve all of its objectives, the United States must reexamine long-held assumptions and policy choices. Without doing so, the costs of maintaining increased outreach to and engagement with Indo-Asia-Pacific countries outside of Northeast Asia will soon become unsustainable. The tyranny of distance in the Pacific theater and the costs associated with sending a company of Soldiers from Hawaii, Alaska, or Washington State to Australia make this point all the more stark for the land-based U.S. Army.

The Army's enduring presence in Northeast Asia appears to be one of those long-held policy choices ripe for a reassessment. The preceding pages do not suggest completely abandoning a forward presence in Northeast Asia—to the contrary, deterrence through punishment remains a key, vital element in U.S. policy toward the erratic regime in Pyongyang, made tangible by the presence of American boots on the ground in South Korea. Nonetheless, it seems plausible that a reconfigured U.S. Army presence in the Indo-Asia-Pacific region would enable the United States to achieve its objectives in the theater more effectively and more efficiently, reducing costs to the military, while simultaneously expanding the array of objectives the forward-based forces can help to achieve.

Specifically, by reorienting some of its existing forward presence from Northeast Asia toward Southeast Asia, the U.S. Army could make its efforts at promoting, enhancing, opening, and sustaining key relationships cheaper and easier to fulfill. At the same time, the U.S. Army could continue to play its critical role in assuring treaty allies and partners throughout the entire theater of the steadfastness of America's commitments.

ENDNOTES

1. U.S. Department of Defense, *Sustaining U.S. Global Leadership: Priorities for 21st Century Defense*, commonly known as the Defense Strategic Guidance, Washington, DC: DoD, January 2012, p. 2.

2. In contrast, U.S. European Command has been led by Army, Navy, Air Force, and Marine Corps officers; U.S. Central Command by Army, Navy, and Marine Corps officers; and U.S. Southern Command by Army, Navy, Air Force, and Marine Corps officers.

3. "U.S. Force Posture Strategy in the Asia Pacific Region: An Independent Assessment," Washington, DC: Center for Strategic and International Studies, August 2012; Thomas G. Mahnken *et al.*, "Asia in the Balance: Transforming U.S. Military Strategy in Asia," Washington, DC: American Enterprise Institute, June 2012; and Lynn E. Davis *et al.*, "U.S. Overseas Military Presence: What Are the Strategic Choices?" Santa Monica, CA: RAND, 2012. See also the 2001 *Quadrennial Defense Review* (QDR) report, which called for increased U.S. Navy presence in the Western Pacific for the U.S. Air Force to develop contingency basing plans in the Pacific and Indian Oceans. U.S. Department of Defense, *Quadrennial Defense Review*, Washington, DC: DoD, September 30, 2001, available from *www.defense.gov/pubs/qdr2001.pdf*.

4. See for example, U.S. Department of Defense, *Strengthening U.S. Global Defense Posture Report To Congress*, Washington, DC: DoD, September 2004.

5. Gordon Adams, "Fiscal Cliff Notes," *Foreign Policy*, November 6, 2012, available from *www.foreignpolicy.com/articles/2012/11/06/fiscal_cliff_notes?page=0,0*. Adams cites acquisition policy, military overhead or "the back office," and personnel policy as three areas most in need of scrutiny.

6. Leon E. Panetta, speech delivered at Georgetown University, February 6, 2013, available from *www.defense.gov/Speeches/Speech.aspx?SpeechID=1747*.

7. Marcus Weisgerber, "Pentagon Projects $35B O&M Shortfall," *Defense News*, March 5, 2013, available from *www.defensenews.com/article/20130305/DEFREG02/303050015/Pentagon-Projects-35B-O-M-Shortfall?odyssey=nav|head*.

8. Cheryl Pellerin, "DOD Comptroller: Sequestration devastates U.S. military readiness," *American Forces Press Service*, May 10, 2013, available from *www.af.mil/news/story.asp?id=123348049*;

Seth Robson and Jon Rabiroff, "Cuts would affect readiness in Pacific, military leaders say," *Stars and Stripes*, February 28, 2013, available from *www.stripes.com/news/sequestration/cuts-would-affect-readiness-in-pacific-military-leaders-say-1.210012*.

9. Jeanette Steele, "SD Destroyer, Frigate Won't Deploy," *The San Diego Union-Tribune*, March 6, 2013, available from *www.utsandiego.com/news/2013/mar/06/preble-rentz-sequestration/*.

10. Interview with a senior field grade officer on the Army staff, February 20, 2013.

11. U.S. Air Force, March 17, 2012, as cited in Peter W. Singer, "Separating Sequestration Facts from Fiction," Washington, DC: The Brookings Institution, September 23, 2012, available from *www.brookings.edu/~/media/research/files/articles/2012/9/sequestration%20singer/sequestration%20singer.pdf*.

12. Office of the Under Secretary of Defense (Comptroller), "National Defense Budget Estimates for FY 2014," Washington, DC: Office of the Comptroller, May 2013, available from *comptroller.defense.gov/defbudget/fy2014/FY14_Green_Book.pdf*.

13. Then Secretary of Defense Robert Gates famously noted in February 2011, "In my opinion, any future defense secretary who advises the president to again send a big American land army into Asia or into the Middle East or Africa should 'have his head examined.'" Thom Shanker, "Warning Against Wars Like Iraq and Afghanistan," *The New York Times*, February 25, 2011, available from *www.nytimes.com/2011/02/26/world/26gates.html*.

14. U.S. Department of Defense, "Sustaining U.S. Global Leadership: Priorities for 21st Century Defense," commonly known as the Defense Strategic Guidance, January 2012, p. 4. Emphasis added.

15. Additionally, there is some evidence of an ongoing debate within the U.S. military over its role in so-called "Phase 0" operations—those that are meant to shape the security environment and prevent conflict before it can happen, such as security cooperation events.

16. See, for example, James Dobbins, "Learning Curve: 'Never Again' is the wrong lesson to draw from the Iraq war," *Foreign Policy*, March 13, 2013, available from *www.foreignpolicy.com/articles/2013/03/13/learning_curve*; and Daniel Byman and Renanah Miles, "A Modest Post-Assad Plan" *The National Interest*, November-December 2012, available from *nationalinterest.org/article/modest-post-assad-plan-7624?page=3*.

17. See, for example, "Developing East Asia slows, but continues to lead global growth at 7.1% in 2013," *World Bank* press release, October 13, 2013; Lucas Kawa, "The 20 Fastest Growing Economies In The World," *Business Insider*, October 24, 2012; and "World Economic Outlook Update," *International Monetary Fund*, July 9, 2013.

18. "Top Trading Partners—Total Trade, Exports, Imports, Year-to-Date December 2012," Washington, DC: U.S. Census Bureau, available from *www.census.gov/foreign-trade/statistics/highlights/top/top1212yr.html*.

19. "Top 25 U.S. International Merchandise Trade Partners by Value: 1970—2001," Washington, DC: U.S. Bureau of Transportation Statistics, available from *apps.bts.gov/publications/us_international_trade_and_freight_transportation_trends/2003/html/table_05.html*.

20. *Ibid.*

21. For example, see Donald L. Berlin, "Neglected no longer: strategic rivalry in the Indian Ocean," *The Harvard International Review*, Vol. 24, No. 2, pp. 26-31; Robert D. Kaplan, "Center stage for the twenty-first century," *Foreign Affairs*, Vol. 88, No. 2, pp. 16-32; David Michel and Russell Sticklor, "Indian Ocean Rising: Maritime and Security Policy Challenges," David Michel and Russell Sticklor, eds., *Indian Ocean Rising: Maritime Security and Policy Challenges*, Washington, DC: The Stimson Center, 2012, pp. 9-22; and the prepared statement of Admiral Samuel J. Locklear, commander of U.S. Pacific Command, before the House Armed Services Committee on March 5, 2013, available from *docs.house.gov/meetings/AS/AS00/20130305/100393/HHRG-113-AS00-Wstate-LocklearUSNA-20130305.pdf*.

22. Christian Bouchard and William Crumplin, "Neglected no longer: the Indian Ocean at the forefront of world geopolitics and global geostrategy," *Journal of the Indian Ocean Region*, Vol. 6, No. 1, June 2010, pp. 26- 51.

23. Australian Government, "Australia in the Asian Century," October 2012, p. 74, available from *asiancentury.dpmc.gov.au/*.

24. Declan Walsh, "Chinese Company Will Run Strategic Pakistani Port," *The New York Times*, January 31, 2013, available from *www.nytimes.com/2013/02/01/world/asia/chinese-firm-will-run-strategic-pakistani-port-at-gwadar.html?_r=0*. Jamil Anderlini and Gwen Robinson, "China-Myanmar pipeline to open in May," *Financial Times*, January 21, 2013, available from *www.ft.com/cms/s/0/faf733ae-63b6-11e2-af8c-00144feab49a.html#axzz2Oku66wep*.

25. See, for example, "Global Trends 2025: A Transformed World," Washington, DC: National Intelligence Council p. 7, available from *www.dni.gov/files/documents/Global%20Trends_2025%20Report.pdf*; Clyde Prestowitz, *Three Billion New Capitalists: The Great Shift of Wealth and Power to the East*, New York, NY: Basic Books, 2005; Raymond J. Ahearn, "Rising Economic Powers and the Global Economy: Trends and Issues for Congress," Congressional Research Service Report, Washington, DC: Congressional Research Service, August 22, 2011; and "Global wealth shifting to Asia, says former World Bank chief," *ABC Radio Australia*, March 23, 2012, available from *abcasiapacificnews.com/stories/201203/3461603.htm*.

26. Special Operations Forces are a major exception to this. They are slated to grow in the coming years.

27. David W. Barno, Nora Bensahel, and Travis Sharp argue that the Army should consist of anywhere from 482,000 to 430,000 Active-Duty Soldiers, depending on the particular strategy chosen and the amount of budget savings desired. See "Hard Choices: Responsible Defense in an Age of Austerity," Washington, DC: Center for a New American Security, October 2011, available from *www.cnas.org/files/documents/publications/CNAS_Hard-Choices_BarnoBensahelSharp_0.pdf*. More recently, Gary Roughead and Kori Schake argue that an Army Active Duty end strength of 290,000, consistent with a strategy that shifts emphasis away from long ground wars and toward rapid response and greater reliance

on allies, should be considered. See "National Defense in a Time of Change," Washington, DC: The Brookings Institution, February 2013, available from *www.brookings.edu/~/media/research/files/ papers/2013/02/us%20national%20defense%20changes/thp_roughead-discpaper.pdf*.

28. Lance M. Bacon, "Cutting 31,000 more soldiers: How small will the Army get?" *Army Times*, March 5, 2013.

29. Chris Carroll, "Army, Air Force chiefs defend cuts in end strength," *Stars and Stripes*, January 27, 2012, available from *www.stripes.com/news/army-air-force-chiefs-defend-cuts-in-end-strength-1.166957*.

30. Respectively, the North Atlantic Treaty Organization, the South East Asia Treaty Organization, and the Central Treaty Organization.

31. James Kitfield, "Is Obama's 'Pivot to Asia' Really a Hedge Against China?" *The Atlantic*, June 8, 2012, available from *www. theatlantic.com/international/archive/2012/06/is-obamas-pivot-to-asia-really-a-hedge-against-china/258279/*.

32. As quoted in Howard LaFranchi, "US 'pivot to Asia': Is John Kerry retooling it?" *Christian Science Monitor*, February 20, 2013, available from *www.csmonitor.com/USA/Foreign-Policy/2013/0220/US-pivot-to-Asia-Is-John-Kerry-retooling-it*.

33. "US re-focussing its Asia 'pivot'," *Bangkok Post*, March 5, 2013, available from *www.bangkokpost.com/news/asia/338860/us-still-intent-on-new-asia-focus-state-department*.

34. For more on how the U.S. Army might contribute to security and confidence building measures with China, see John R. Deni, "Strategic Landpower and the U.S. Army Role in the Indo-Asia-Pacific," *Parameters*, Vol. 43, No. 3, Autumn 2013.

35. Matt Siegel, "As Part of Pact, U.S. Marines Arrive in Australia, in China's Strategic Backyard," *The New York Times*, April 4, 2012, available from *www.nytimes.com/2012/04/05/world/ asia/us-marines-arrive-darwin-australia.html*; Gidget Fuentes, "Marines pave way for larger force in Australia," *Marine Corps Times*,

October 16, 2012, available from *www.marinecorpstimes.com/ news/2012/10/marine-australia-darwin-success-101612/*.

36. Such a large rotation is only likely in the final year — 2016 — of the initial 5-year agreement.

37. Matt Siegel, "As Part of Pact, U.S. Marines Arrive in Australia, in China's Strategic Backyard," *The New York Times*, April 4, 2012, available from *www.nytimes.com/2012/04/05/world/asia/us-marines-arrive-darwin-australia.html*.

38. Interview with civilian officials from U.S. Army Pacific, October 11, 2013.

39. Choe Sang-Hun, "North Korea Declares 1953 War Truce Nullified," *The New York Times*, March 11, 2013, available from *www.nytimes.com/2013/03/12/world/asia/north-korea-says-it-has-nullified-1953-korean-war-armistice.html?_r=0*.

40. A different subject for a different study, but suffice it to say that this monograph assumes it is in Washington's interests to have other countries seeking to gain American favor.

41. Carl Munoz, "The Philippines re-opens military bases to US forces," *The Hill*, June 6, 2012, available from *thehill.com/blogs/defcon-hill/operations/231257-philippines-re-opens-military-bases-to-us-forces-*; Alexis Romero, "US-Phl agree on close defense cooperation," *The Philippine Star*, March 19, 2013, available from *www. philstar.com/headlines/2013/03/19/921621/us-phl-agree-close-defense-cooperation*.

42. Nikko Dizon, "PH: US, allies may use military bases," *Philippine Daily Inquirer*, June 28, 2013, available from *globalnation. inquirer.net/78885/ph-us-allies-may-use-military-bases*.

43. Manuel Mogato, "Manila plans air, naval bases at Subic with access for U.S., officials say," *Reuters*, June 27, 2013, available from *www.reuters.com/article/2013/06/27/us-philippines-usa-idUSBRE95Q0C120130627*; see also "New Philippine base will send message to China — Expert," *ABS-CBNnews.com*, June 27, 2013, available from *www.abs-cbnnews.com/focus/06/27/13/new-philippine-base-will-send-message-china-expert*.

44. Richard Weitz, "Nervous Neighbors: China Finds a Sphere of Influence," *World Affairs Journal*, March/April 2011. Weitz writes, "Unsurprisingly, it has been the Vietnamese who have most eagerly sought to work with their former adversary to balance the regional colossus." See also Nayan Chanda, "The slow rapprochement," *The American Review*, November 2012, available from *americanreviewmag.com/stories/The-slow-rapprochement*. Admittedly though, there is not universal agreement on the forces motivating the Vietnamese government today. See Carlyle Thayer, "Strategic Posture Review: Vietnam," *World Politics Review*, January 15, 2013, pp. 1-11; as well as Carlyle Thayer, "Vietnam's Strategic Outlook," *Security Outlook of the Asia Pacific Countries and Its Implications for the Defense Sector*, NIDS Joint Research Series No. 7, The NIDS International Workshop on Asia Pacific Security, Tokyo, Japan: The National Institute for Defense Studies Japan, 2013, pp. 69-88.

45. Il Hyun Cho and Seo-Hyun Park, "The Rise of China and Varying Sentiments in Southeast Asia toward Great Powers," *Strategic Studies Quarterly*, Vol. 7, Issue 2, Summer 2013.

46. Previously referred to as Task Force East, the U.S. forward operating sites in Romania and Bulgaria have collectively been called the Black Sea - Area Support Team (BS-AST) since 2011.

47. Seth Robson, "New bases in Bulgaria, Romania cost U.S. over $100M," *Stars and Stripes*, October 17, 2009, available from *www.stripes.com/news/new-bases-in-bulgaria-romania-cost-u-s-over-100m-1.95658*.

48. This figure is an estimate, based on validated requirements. The $11M figure covers warm basing of the Temporary Food Event facility in Romania at Mihail Kogalniceanu Air Base—meaning it is able to support an announced unit mission within 30 days of notification—and the cold basing of the TFE facility in Bulgaria at Novo Selo—meaning it is able to support to an announced unit mission within 60-90 days of notification.

49. Michael Lostumbo *et al.*, *Overseas Basing of U.S. Military Forces An Assessment of Relative Costs and Strategic Benefits*, Santa Monica, CA: RAND, 2013, p. 177.

50. *Ibid.*, p. 176.

51. *Ibid.*

52. U.S. Department of Defense, *Statistical Compendium On Allied Contributions To The Common Defense*, Washington, DC: DoD, 2004, p. B-21.

53. *Ibid.*, p. B-7.

54. *Ibid.*, p. B-16.

55. Ely Ratner, "Rebalancing to Asia with an Insecure China," *The Washington Quarterly*, Vol. 36, No. 2, Spring 2013, pp. 21-38.

56. Kenneth Lieberthal and Wang Jisi, *Addressing U.S.-China Strategic Distrust*, Washington, DC: The Brookings Institution, 2012, p. 7; Andrew J. Nathan and Andrew Scobell, "How China Sees America: The Sum of Beijing's Fears," *Foreign Affairs*, Vol. 91, No. 5, September/October 2012, pp. 32-47.

57. Lieberthal and Jisi, p. 11.

58. Nathan and Scobell, pp. 32-47; Robert S. Ross, "The Problem With the Pivot: Obama's New Asia Policy Is Unnecessary and Counterproductive," *Foreign Affairs*, Vol. 91, No. 6, November/December 2012, pp. 70-82.

59. Cecil Morella, "Philippines boosts military to resist 'bullies'," *Agence France-Presse*, May 21, 2013, available from *www.google.com/hostednews/afp/article/ALeqM5jeDtBpfWgBRcqS1PdWv2-M2vRCQg?docId=CNG.2083e68bf6225d54f1ac44ad4d6d7e4c.291*.

60. James A. Baker, *The Politics of Diplomacy: Revolution, War, and Peace, 1989-1992*, New York: Putnam's Sons, 1995.

61. The analogy to Germany is not perfect, as obviously there are no U.S. troops stationed in China. However, U.S. troops' presence in the Indo-Asia-Pacific clearly has a reassuring affect upon American allies. This should not be lost on Chinese leadership.

62. Ratner, p. 23.

63. "A neutral US helpful to stability in S China Sea," *China Daily USA*, May 7, 2012, available from *usa.chinadaily.com.cn/opinion/2012-05/07/content_15226749.htm*.

64. Ratner, p. 28.

65. Peter Lavoy, testimony delivered before the House Armed Services Committee, Washington, DC: U.S. House of Representatives, March 28, 2012, available from *armedservices.house.gov/index.cfm/files/serve?File_id=4fa3e4af-2edf-4152-adf9-97a6fbc0176e*.

66. "The New Korea," *U.S. Forces Korea*, October 2010, p. 27, available from *www.usfk.mil/usfk/Uploads/120/USFK_SD_SPREAD_10MB.pdf*.

67. Jack Kim and Louis Charbonneau, "South Korea says to strike back at North if attacked," *Reuters*, March 6, 2013, available from *www.reuters.com/article/2013/03/06/us-korea-north-idUS-BRE92508720130306*.

68. "The New Korea."

69. Bruce W. Bennet, *A Brief Analysis of the Republic of Korea's Defense Reform Plan*, Santa Monica, CA: RAND, 2006. South Korea is also wrestling with the long-term implications of an aging population and one of the world's lowest birth rates.

70. As part of then-Secretary of Defense Donald Rumsfeld's *Global Posture Review*, the United States reduced its total troop presence in South Korea from 35,000 troops to 28,000 by withdrawing one U.S. Army brigade combat team and two attack helicopter battalions.

71. Deterrence by punishment works by assuring the aggressor that he will have to endure unacceptable costs in the event of aggression. Deterrence by denial works by assuring the aggressor that defenses are **so** effective, aggression in the first place would be futile. Even though the United States is in the process of moving most of its forces south of the demilitarized zone, it is unlikely in the extreme that a massive North Korea invasion of the South would **not** entail at least some American casualties, even if the South Korean military bore the brunt of the attack.

72. Richard L. Armitage and Joseph S. Nye, "The U.S.-Japan Alliance: Anchoring Stability in Asia," Washington, DC: Center for Strategic and International Studies, August 2012, p. 2, available from *csis.org/files/publication/120810_Armitage_US JapanAlliance_Web.pdf*.

73. *Ibid.*, p. 12.

74. Christopher W. Hughes, "Japan, Ballistic Missile Defence and Remilitarisation," *Space Policy*, Vol. 29, Issue 2, May 2013, p. 129; Richard J. Samuels, "Japan's Goldilocks Strategy," *The Washington Quarterly*, Vol. 29, No. 4, Autumn 2006, p. 115.

75. These figures do not include the nearly 6,600 Sailors and nearly 2,000 Marines afloat in the region. See *www.globalsecurity. org/military/library/report/2011/hst1103.pdf*.

76. Nevertheless, the U.S. Army is assigned as Department of Defense Executive Agent for several functions worldwide, including mortuary affairs, chemical and biological defense, and military postal services. In some cases, such as postal services, the Army has primary responsibility for management; in others, such as mortuary affairs, the Army is responsible for support to Army units and backup general support to the other military services, as required. For a list of all Army executive agent mission areas, see *www.oaa.army.mil/AEA_functions.aspx*.

77. For budgetary reasons related to the strategic context outlined previously, the addition of **more** Soldiers to the theater, in addition to those currently forward based in the region, is not considered here.

78. United States Army Pacific, "Partnering in the Pacific Theater: Assuring Security and Stability Through Strong Army Partnerships," Schofield Barracks, HI: USARPAC, April 26, 2012.

79. Karen Parrish, "South Korea Mission Strategically Important, Officials Say," *American Forces Press Service*, March 28, 2012, available from *www.defense.gov/news/newsarticle.aspx?id=67742*.

80. Jim Garamone, "Commander Seeks Enhanced Deterrent on Korean Peninsula," *American Forces Press Service*, June

12, 2012, available from *www.defense.gov/News/NewsArticle. aspx?ID=116710.*

81. In fact, the Australians appear interested in even closer cooperation with the U.S. Army—witness the assignment of a two-star Australian general as a deputy commander within U.S. Army Pacific. Audrey Mcavoy, "Aussie 2-star gets senior post at USARPAC," *Associated Press,* Monday, August 20, 2012.

82. Key U.S. national security documents, such as the 2010 QDR report, spell out explicitly the necessity of building coalitions to achieve common objectives. Meanwhile, the Lowy Institute, an independent public policy think tank based in Sydney, found that 82 percent of Australians support the alliance with the United States. See Alex Oliver, *Australia and the World: Public Opinion and Foreign Policy,* Sydney, Australia: The Lowy Institute for International Policy, 2013, available from *www.lowyinstitute. org/publications/lowy-institute-poll-2013.*

83. John R. Deni, "Whose Responsibility is Interoperability?" *Small Wars Journal,* June 26, 2013, available from *smallwarsjournal. com/jrnl/art/whose-responsibility-is-interoperability.*

84. Leon Panetta, Remarks at the Institute for Defence Studies and Analyses in New Delhi, India, June 6, 2012, available from *www.defense.gov/transcripts/transcript.aspx?transcriptid=5054.*

85. *Ibid.*

86. Interview with civilian officials from U.S. Army Pacific, October 11, 2013.

87. See, for example, John R. Deni, *The Future of American Landpower: Does Forward Presence Still Matter? The Case of the Army in Europe,"* Carlisle, PA: Strategic Studies Institute, U.S. Army War College, October 9, 2012, available from *www.strategic studiesinstitute.army.mil/pubs/display.cfm?pubID=1130;* Lostumbo et al., p. 95.

88. Oliver, p. 8.

89. Manuel Mogato, "U.S. military to boost Philippines presence; China tells army to be prepared," *Reuters*, December 12, 2012, available from *www.reuters.com/article/2012/12/12/us-philippines-usa-idUSBRE8BB0LL20121212*.

90. Section 25 of the Constitution of the Philippines requires that all foreign military bases, troops, or facilities shall not be allowed in the Philippines except under a treaty approved by the Filipino Senate and, when required by the Filipino Congress, ratified by a majority of citizens in a national referendum.

91. Mogato, "U.S. military to boost Philippines presence; China tells army to be prepared."

92. Shawn W. Crispin, "When allies drift apart," *Asia Times Online*, February 14, 2009, available from *www.atimes.com/atimes/Southeast_Asia/KB14Ae01.html*.

93. Emma Chanlett-Avery and Ben Dolven, "Thailand: Background and U.S. Relations," Congressional Research Service Report, Washington, DC: Congressional Research Service, June 5, 2012.

94. Crispin, "When allies drift apart."